D0947230

H.S.M. Coxeter
P. Du Val
H.T. Flather
J.F. Petrie

The Fifty-Nine Icosahedra

With 20 Plates and 9 Figures

Springer-Verlag
New York Heidelberg Berlin

AMS Subject Classifications: 51M20, 52A25

Library of Congress Cataloging in Publication Data
Coxeter, H. S. M. (Harold Scott Macdonald), 1907–
The fifty-nine icosahedra.
Originally published: Toronto: University of Toronto Press, 1938. (University of Toronto studies. Mathematical series; no. 6)
1. Icosahedra. I. Du Val, Patrick, 1903– . II. Title. III. Title: The 59 icosahedra. IV. Series: University of Toronto studies. Mathematical series; no. 6.
QA491.C6 1982 516′.15 82-10582

The first printing was published in 1938 by The University of Toronto Press.

Printed and bound by R. R. Donnelley & Sons, Harrisonburg, VA.
Printed in the United States of America.

9 8 7 6 5 4 3 2 1

ISBN 0–387–**90770–X** Springer-Verlag New York Heidelberg Berlin
ISBN 3–540–**90770–X** Springer-Verlag Berlin Heidelberg New York

PREFACE TO THE SPRINGER EDITION

The Fifty-Nine Icosahedra was originally published in 1938 as No. 6 of "University of Toronto Studies (Mathematical Series)". Of the four authors, only Coxeter and myself are still alive, and we two are the authors of the whole text of the book, in which any signs of immaturity may perhaps be regarded leniently on noting that both of us were still in our twenties when it was written. Neither of the others was a professional mathematician. Flather died about 1950, and Petrie, tragically, in a road accident in 1972. Petrie's part in the book consisted in the extremely difficult drawings which constitute the left half of each of the plates (the much simpler ones on the right being mine). A brief biographical note on Petrie will be found on p. 32 of Coxeter's *Regular Polytopes* (3rd. ed., Dover, New York, 1973); and it may be added that he was still a schoolboy when he discovered the regular skew polygons that are named after him, and are the occasion for the note on him in Coxeter's book. (Coxeter also was a schoolboy when some of the results for which he will be most remembered were obtained; he and Petrie were schoolboy friends and used to work together on polyhedron and polytope theory.)

Flather's part in the book consisted in making a very beautiful set of miniature models of all the fifty-nine figures. These are still in existence, and in excellent preservation. They can be seen by anyone who is interested, on making written application for an appointment to the Head of the Department of Pure Mathematics and Mathematical Statistics, University of Cambridge, 16 Mill Lane, Cambridge CB2 1SB, England, where they are permanently housed.

Apart from partial precursors referred to in Section 1 of the present book, Coxeter was the first to devote serious study to stellations of the icosahedron. He was certainly the first to enumerate them completely, using the arguments and notation of Section 2 of the book, some years before its original publication (those in Section 3 being mine). In the actual definition, some part was played by J. C. P. Miller, as indicated at the end of Section 1.

Cambridge, 1982 PATRICK DU VAL

CONTENTS

The
Fifty-Nine
Icosahedra

THE FIFTY-NINE ICOSAHEDRA

1. Introduction

In this paper we enumerate and describe the polyhedra that can be derived from the five Platonic solids by *stellation*, *i.e.*, by extending or "producing" the faces until they meet again, always preserving the rotational symmetry of the original solid. For the faces to meet again, the dihedral angles between them must be obtuse; therefore the tetrahedron and cube cannot be stellated. About the year 1619, Kepler* stellated the octahedron and dodecahedron, obtaining the *stella octangula* and the "small" and "great" *stellated dodecahedra*. The stella octangula may be called a *compound* polyhedron, since it takes the form of two tetrahedra so placed that their eight vertices are the vertices of a cube. It is evident that no further stellation of the octahedron is possible. The stellated dodecahedra are *regular* polyhedra, under the slightly extended definition which admits the *pentagram* (or star pentagon) as a regular polygon; their faces are pentagrams, and their vertices are uniformly and regularly surrounded.

In 1809, Poinsot† discovered the reciprocals of Kepler's star dodecahedra: the *great dodecahedron* (dodecaèdre de la troisième espèce) and the *great icosahedron* (icosaèdre de la septième espèce). Two years later, Cauchy† showed that these arise by stellating the dodecahedron and icosahedron, respectively, and proved that they exhaust the list of finite regular polyhedra. It is an immediate consequence of his constructions, that every stellated dodecahedron (preserving symmetry) is regular. Thus the only Platonic solid whose stellations remained to be investigated was the icosahedron.

---

*J. Kepler, *Opera Omnia*, vol. V (1864): "Harmonia Mundi", p. 122 (XXVI. Propositio).

†Poinsot, Cauchy, Bertrand and Cayley, *Abhandlungen über die regelmässigen Sternkörper* (Leipzig, 1906), pp. 38, 35, 56-62.

Just as a tetrahedron can be inscribed in a cube, so a cube can be inscribed in a dodecahedron.* By reciprocation, this leads to an octahedron circumscribed about an icosahedron.† In fact, each of the twelve vertices of the icosahedron divides an edge of the octahedron according to the "golden section". Given the icosahedron, the circumscribed octahedron can be chosen in five ways, giving a compound of five octahedra‡ which comes under our definition of a stellated icosahedron. (The reciprocal compound, of five cubes whose vertices belong to a dodecahedron, is a stellated triacontahedron, and so does not concern us here, since the triacontahedron is only semi-regular.) Another stellated icosahedron can at once be deduced, by stellating each octahedron into a stella octangula, thus forming altogether a compound of ten tetrahedra. Further, we can choose one tetrahedron from each stella octangula, so as to derive a compound of five tetrahedra, which still has all the rotational symmetry of the icosahedron (*i.e.*, the icosahedral group), although it has lost the reflections. By reflecting this figure in any plane of symmetry of the icosahedron, we obtain the complementary set of five tetrahedra. These two sets of five tetrahedra are enantiomorphous, *i.e.*, not directly congruent, but related like a pair of shoes. A figure which possesses at least one plane of symmetry (so that it is directly congruent to its mirror-image) is said to be *reflexible*; a figure (such as the compound of five tetrahedra) which possesses no plane of symmetry (so that it is enantiomorphous to its mirror-image) is said to be *chiral*.

All these figures were thoroughly described and elegantly photographed by M. BRÜCKNER, who discovered a number of new stellations of the icosahedron. Several more are due to A. H. WHEELER,§ for whose inspiration we would express our gratitude. The following table shows our symbols for Wheeler's stellated icosahedra, and the location of Brückner's pictures of those that were known to him.

---

*Kepler, *loc. cit.*, p. 271 (Fig. 38).

†Poinsot, Cauchy, Bertrand and Cayley, *loc. cit.*, p. 61.

‡M. Brückner, *Vielecke und Vielflache* (Leipzig, 1900), p. 206.

§"Certain forms of the icosahedron, and a method for deriving and designating higher polyhedra", *Proc. Internat. Math. Congress*, Toronto, 1924, Vol. 1, pp. 701-708.

		Wheeler	Brückner
A	(the icosahedron itself)	1	(not shown)
B	(a "triakisicosahedron")	2	Fig. 2, Taf. VIII
C	(the five octahedra)	3	Fig. 6, Taf. IX
D		4	Fig. 17, Taf. IX
$\mathbf{De_2}$		5	
$\mathbf{E}f_1$	(the five tetrahedra, *laevo*)	6	(not shown)
$\mathbf{Ef_1}$	(the five tetrahedra, *dextro*)	7	Fig. 11, Taf. IX
$\mathbf{Ef_1}$	(the ten tetrahedra)	8	Fig. 3, Taf. IX
$\mathbf{Ef_1g_1}$		9	Fig. 26, Taf. VIII
$\mathbf{De_1}$		10	
G	(the great icosahedron)	11	Fig. 24, Taf. XI
H	(the "complete" stellation)	12	Fig. 14, Taf. XI
$\mathbf{De_2f_2}$		13	
$\mathbf{Fg_2}$		14	
$\mathbf{Ef_1f_2}$	(*dextro*)	15	
$\mathbf{E}f_1\mathbf{f_2}$	(*laevo*)	16	
$\mathbf{Fg_1}$		17	Fig. 3, Taf. X
$\mathbf{Ef_2}$		18	Fig. 20, Taf. IX
F		19	
$\mathbf{e_1f_1g_1}$		20	
$\mathbf{g_1}$		21	
$\mathbf{f_2}$		22	

As might be expected in such a case, an exhaustive enumeration greatly increases the list. We find, altogether, thirty-two reflexible icosahedra (including the ordinary and "great" icosahedra, and the compounds of five octahedra and of ten tetrahedra) and twenty-seven enantiomorphous pairs (including the pair of compounds of five tetrahedra, and Wheeler's Nos. 15, 16). Of the latter, we would draw special attention to the skeletal pair f_1, f_1, either of which may be described as that part of one set of five tetrahedra which is exterior to the other set. For, it will appear that every enantiomorphous pair can be derived from one or two reflexible icosahedra by adding f_1 and f_1, in turn.

The process which we have called stellation has an obvious analogue in two dimensions. The ordinary regular n-gon, $\{n\}$, leads to the stellated n-gons $\{n/d\}$ $(n>4,\ 1<d<\frac{1}{2}n)$. When d is prime to n, $\{n/d\}$ is a single regular star polygon, having n sides which surround the centre d times; *e.g.*, $\{\frac{5}{2}\}$ is the pentagram or five-pointed star. On the other hand, $\{kn/kd\}$ $(k>1)$ is a com-

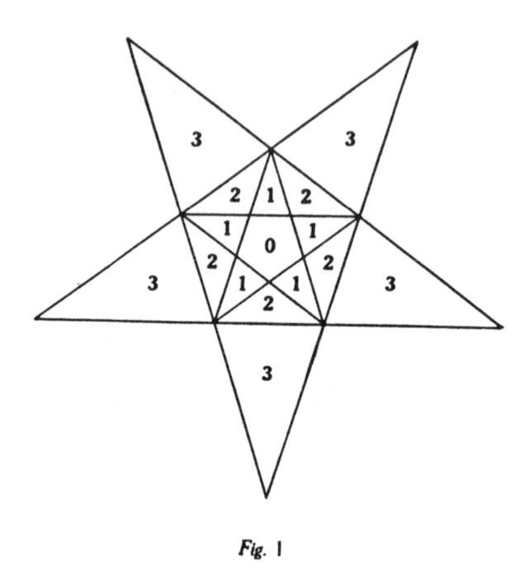

Fig. 1

pound polygon, consisting of k equal and concentric $\{n/d\}$'s with the vertices of a $\{kn\}$; *e.g.*, $\{\frac{6}{2}\}$ is a pair of equilateral triangles with the vertices of a regular hexagon.

The regular dodecahedron is sometimes denoted by $\{5, 3\}$, because it is bounded by pentagons $\{5\}$, which meet by threes at the vertices. If all the faces are stellated so as to become pentagrams $\{\frac{5}{2}\}$, it is easily seen that these pentagrams meet by fives at the new vertices, giving the "small stellated dodecahedron", $\{\frac{5}{2}, 5\}$. When each of these pentagrams has its vertices joined by new edges so as to form a large pentagon, there are still five faces at each vertex, but these go twice round, as at a winding-point on a Riemann surface; so the "great dodecahedron" is denoted by $\{5, \frac{5}{2}\}$. Finally, by stellating these large pentagons into pentagrams, we obtain the "great stellated dodecahedron", $\{\frac{5}{2}, 3\}$. (These names were invented by CAYLEY; the symbols by SCHLÄFLI.)

The ten straight lines in Fig. 1 show how the plane of any one face of $\{5, 3\}$ is met by the planes of ten other faces (namely, all the other faces except the opposite one, which is parallel). These lines divide the plane into a number of regions, of which the finite ones are marked **0**,**1**,**2**,**3**, in a manner that preserves the pentagonal symmetry. We imagine the rest of the twelve planes to be divided and marked the same way. The regions **0** are the faces of the original dodecahedron. The regions **1** are the external ("acces-

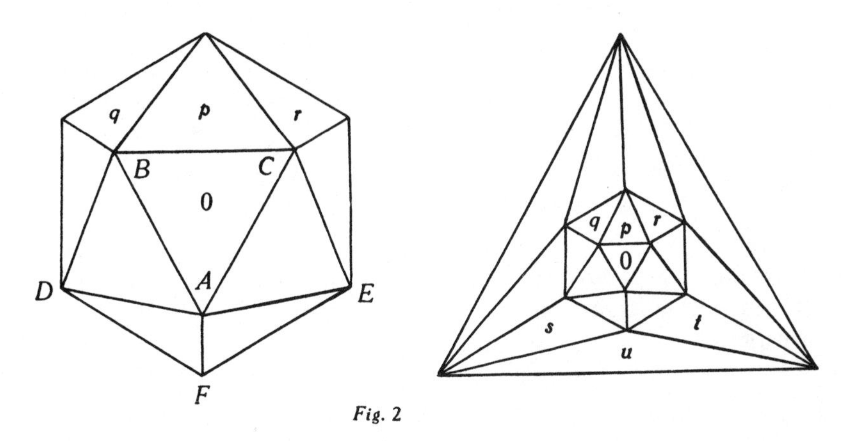

Fig. 2

sible") parts of the faces of $\{\frac{5}{2}, 5\}$. Similarly, the regions **2** and **3** bound $\{5, \frac{5}{2}\}$ and $\{\frac{5}{2}, 3\}$, respectively.

The icosahedron, $\{3, 5\}$, is represented in Fig. 2, first by orthogonal projection and then by a Schlegel diagram. Its stellations are so numerous and complicated that some care is needed in stating precisely what varieties shall be considered properly significant and distinct. J. C. P. MILLER has suggested the following set of five restrictive rules.

(i) The faces must lie in twenty planes, *viz.*, the bounding planes of the regular icosahedron.

(ii) All parts composing the faces must be the same in each plane, although they may be quite disconnected.

(iii) The parts included in any one plane must have trigonal symmetry, with or without reflection. This secures icosahedral symmetry for the whole solid.

(iv) The parts included in any plane must all be "accessible" in the completed solid (*i.e.*, they must be on the "outside". In certain cases we should require models of enormous size in order to see *all* the outside. With a model of ordinary size, some parts of the "outside" could only be explored by a crawling insect).

(v) We exclude from consideration cases where the parts can be divided into two sets, each giving a solid with as much symmetry as the whole figure. But we allow the combination of an enantiomorphous pair having no common part (which actually occurs in just one case).

2. Complete Enumeration of Stellated Icosahedra, by Considering the Possible Faces

The eighteen straight lines in Fig. 3 (analogous to Fig. 1)* show how the plane of any one face **0** of the icosahedron is met by the planes of eighteen other faces (namely, all the other faces except the opposite one). The lines

$$PP', QQ', RR', SS', TT', UU'$$

are typical, in the sense that the rest can be derived from these six by rotations through $\pm\frac{2}{3}\pi$. The faces whose planes make these intersections are marked p, q, r, s, t, u, in Fig. 2. Note that pairs of opposite faces (p, u; q, t; r, s) lead to pairs of parallel lines.

The regions that occur in Fig. 3 are numbered in such a way as to preserve the trigonal symmetry of the face. Enantiomorphous regions, *dextro* and *laevo*, are distinguished by means of Roman and italic type, *e.g.*, 2 and *2*; regions which cannot be separated in this manner are denoted by numbers in clarendon type. We imagine the rest of the twenty faces to be divided and numbered the same way, the rotational sense determined by the distinction between 2 and *2* being the same in every face (as seen from outside). The regions **0** are the faces of the original icosahedron.

*E. Hess, *Über die zugleich gleicheckigen und gleichflächigen Polyeder*, 1876; Brückner, *loc. cit.*, Fig. 17, Taf. II.

PLATE 1

A

0

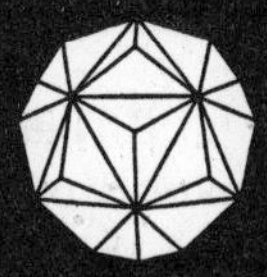

B

1

C

2

D

3 4

Different solids will be obtained by making suitable selections of the other regions; but the rules for selection require careful consideration. From the manner in which we have numbered the regions, the derived solids will all have the *rotational* symmetry of the icosahedron (*i.e.*, they will be transformed into themselves by the operations of the icosahedral group). Whenever every symmetrical region that occurs is accompanied by its enantiomorph (*e.g.*, 2 by *2*), the solid is reflexible and has the *full* symmetry of the icosahedron; in other cases the solid exists in two enantiomorphous forms, which are not "superposable", but are interchanged by reflection in any plane of symmetry of the icosahedron.

It is desirable to see how the line PP' (Fig. 3) behaves when regarded as belonging to the face p (Fig. 2) instead of to the face **0**, to see how QQ' behaves in q, and so on. For this purpose, we ask what symmetries of the icosahedron will transpose **0** and p, **0** and q, etc.

The faces **0** and p are transposed by reflection in the edge BC (or rather, in the plane joining this edge to the centre of the icosahedron); this reflection leaves PP' (which is BC) unchanged, but interchanges *dextro* and *laevo* regions. These same faces can alternatively be transposed by rotation through π about the mid-point of BC (or rather, about the line joining this mid-point to the centre); this rotation preserves the orientation of the regions, but reverses the line PP'.

The faces **0** and q (or **0** and r) are transposed by reflection in the line BD (or CE). Any reflection changes a line of type QQ' into one of type RR'.

The faces **0** and s (or **0** and t) are transposed by rotation through π about the mid-point of AD (or AE). The faces **0** and u are transposed by reflection in DE, or by rotation through π about the mid-point of AF.

Collecting results:

PP' in **0** is of type $P'P$ in p,
QQ' in **0** is of type RR' in q,
RR' in **0** is of type QQ' in r,
SS' in **0** is of type $S'S$ in s,
TT' in **0** is of type $T'T$ in t,
UU' in **0** is of type $U'U$ in u.

PLATE II
E
5 6 7
F
8 9 10
G
11 12

Consider that segment of PP' (Fig. 3) which separates regions marked 5 and 9. Since PP' in **0** is of type $P'P$ in p, this same segment will separate regions *5* and *9* in p. Hence the face of a proper stellated icosahedron must involve either none, two, or all four, of the regions 5,9,*5*,*9*. We shall speak of such a set of regions as a "tetrad".

The next segment of PP', separating regions 4 and **7**, provides still more useful information; for, the corresponding regions in p

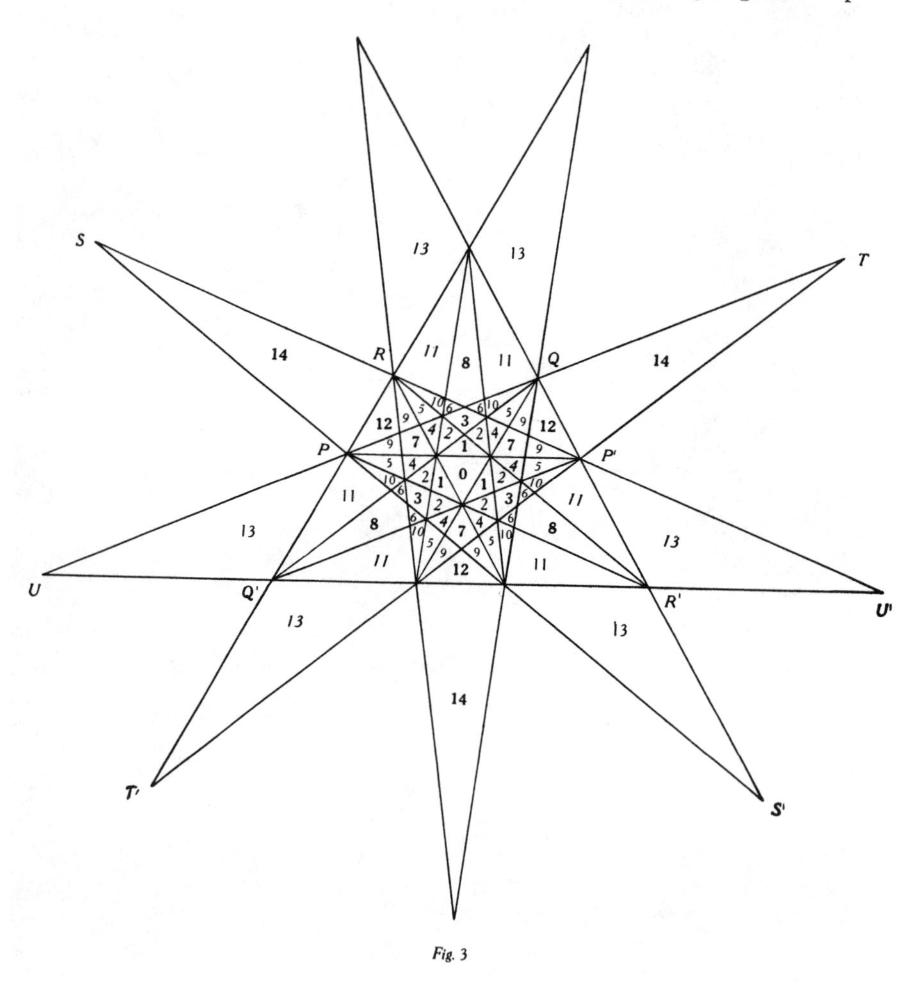

Fig. 3

PLATE III

Face 13
(*See Fig.* 4)

are *4* and (again) **7**. From this we conclude that, whether **7** occurs or not, 4 and *4* can only occur together. Similarly, by considering certain segments of QQ' and RR', we see that 2 and *2* can only occur together, and likewise 11 and *11*. Finally, by considering the outermost segments of SS' and $S'S$, TT' and $T'T$, UU' and $U'U$, we see that 13, *13* and **14** must all occur together (when they occur at all). We can now simplify the numbering of the regions by writing **2** for both 2 and *2*, **4** for 4 and *4*, **11** for 11

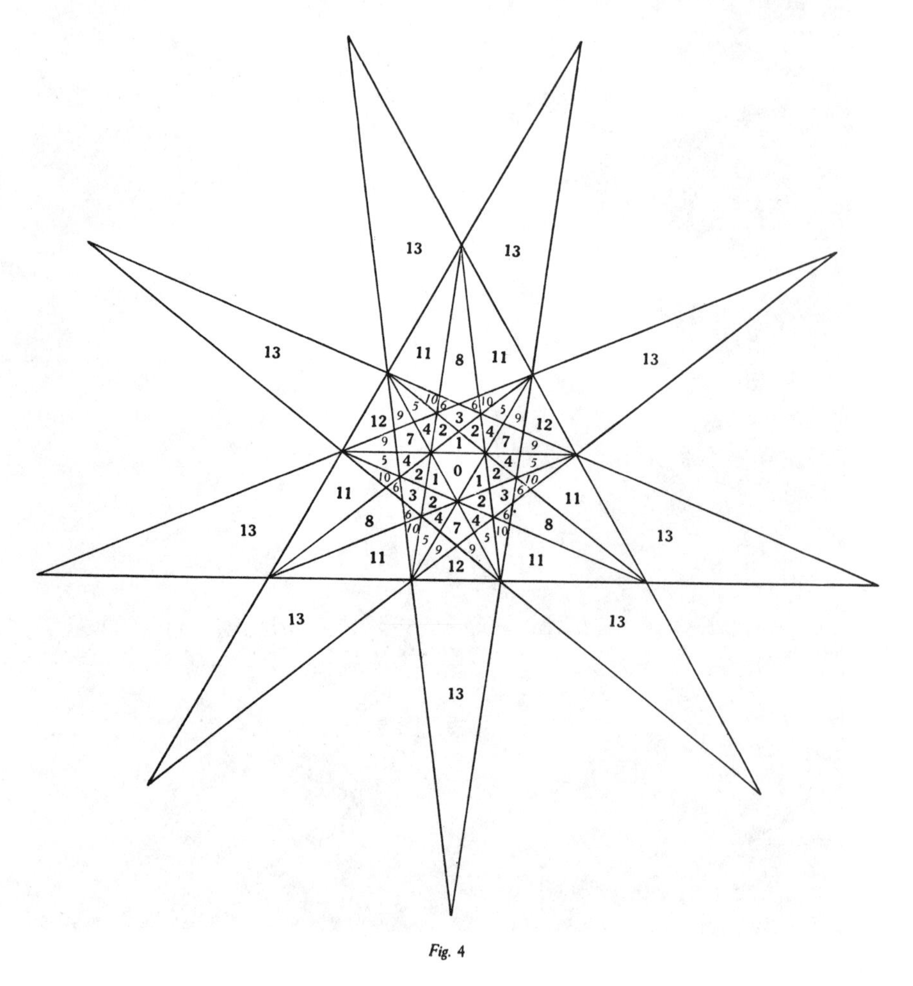

Fig. 4

e_1

3′5

f_1

5′6′9 10

g_1

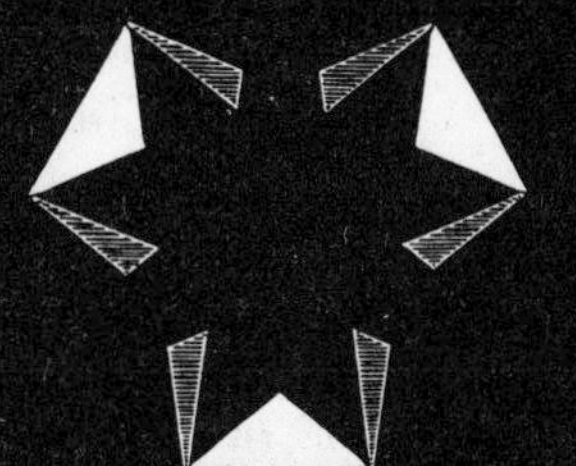

10′12

and *11*, and **13** for all of 13, *13*, **14**. The revised notation is shown in Fig. 4.

By working systematically through the various segments of PP' and $P'P$, QQ' and RR', SS' and $S'S$, TT' and $T'T$, we obtain the "tetrads"

(5, *5*, 9, *9*), (5, *5*, 10, *10*), (6, *6*, 10, *10*),

(**3**, **4**, *5*, *6*), (6, **7**, **8**, *9*), (9, 10, **11**, **12**),

(**3**, **4**, 5, 6), (*6*, **7**, **8**, 9), (*9*, *10*, **11**, **12**).

When regions have been selected so as to bound a solid, the corresponding set of numbers can be used as a symbol for the face of the solid. Thus we say that the original icosahedron has face **0**. The only other faces whose symbols consist of a single number are **1**, **2** and **13**. (These belong to proper solids, since the numbers involved occur in no tetrad.) Of symbols consisting of two numbers, we can certainly have **3 4**, **7 8**, **11 12**, since every tetrad containing **3** contains **4** also, and so on.

In all other cases (by Miller's fifth condition, at the end of § 1), we may only take one of the two numbers **3** and **4**, one of **7** and **8**, one of **11** and **12**. These pairs of numbers are interchangeable; for every face-symbol involving **3** there will be another involving **4** instead, and so on. Accordingly, we introduce letters λ, μ, ν: λ to stand for **3** or **4**, μ for **7** or **8**, ν for **11** or **12**. The last six of our tetrads now become "triads":

(λ, *5*, *6*), (*6*, μ, 9), (9, 10, ν),

(λ, 5, 6), (6, μ, *9*), (*9*, *10*, ν).

It is convenient to represent the eleven symbols λ, 5, *5*, 6, *6*, μ, 9, *9*, 10, *10*, ν by the lines of a certain *graph* (Fig. 5) which looks like a ladder with three rungs. Each "triad" is represented by the lines meeting at one vertex of the graph (other than the end vertices X). The lines that represent the face-symbol of a stellated icosahedron form a single chain, which either is closed, or begins at one of the four vertices X and ends at another. (By making the four points X coincide, we could obtain the edges of a

PLATE V

e_1f_1

3′6′9 10

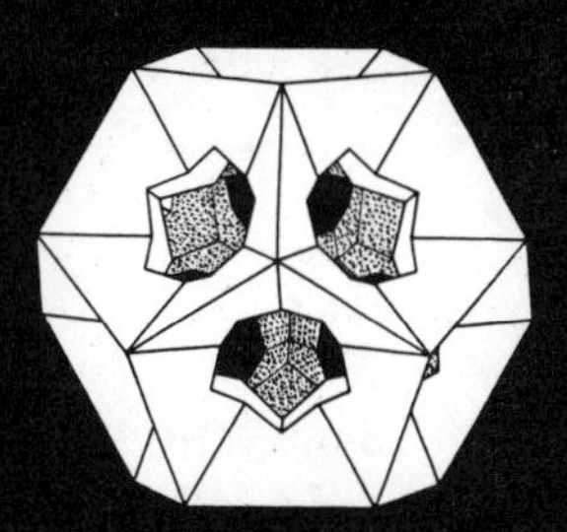

$e_1f_1g_1$

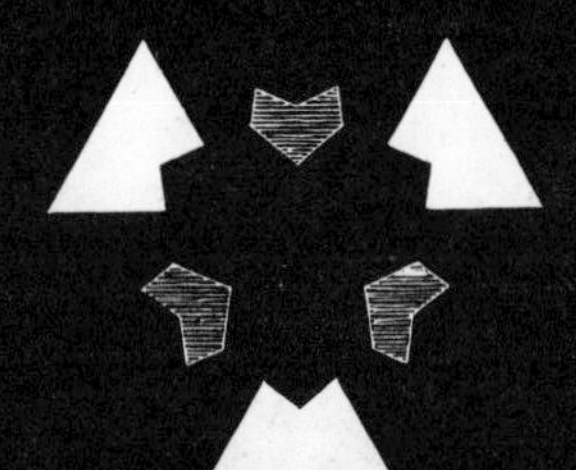

3′6′9 12

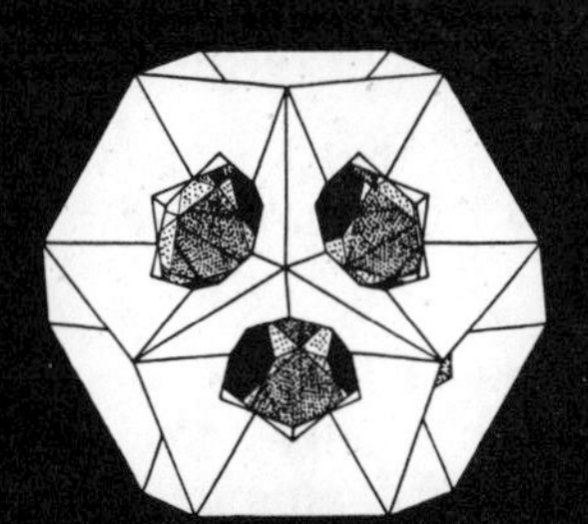

f_1g_1

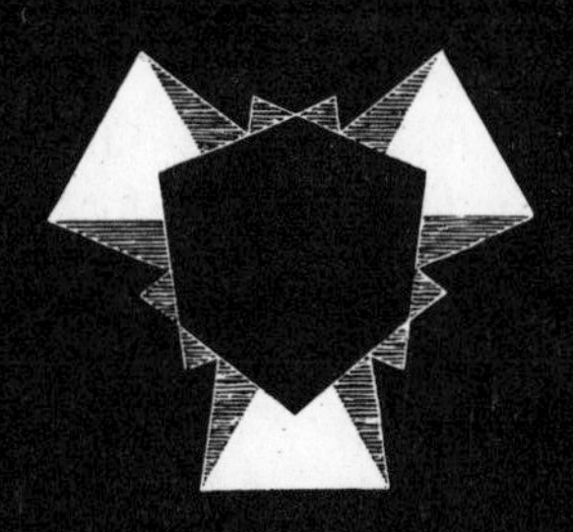

5′6′9 12

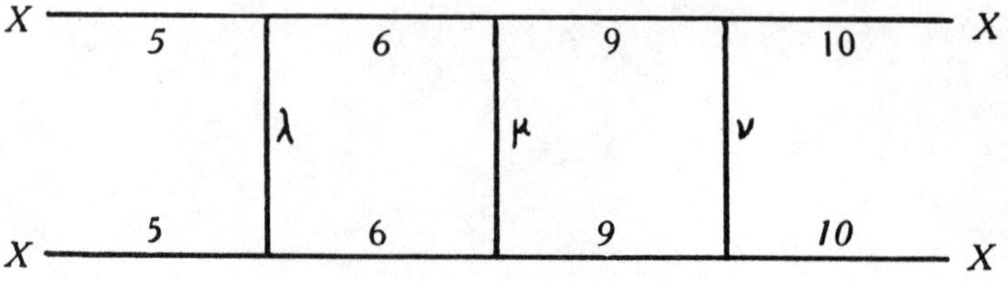

Fig. 5

polyhedron bounded by two triangles and four quadrangles. The sets of lines to be chosen would then be just the closed chains of edges. The point X would represent the tetrad (5, *5*, 10, *10*).) The possible closed chains are

λ 6 μ *6*, λ 6 *9* ν 9 *6*, μ *9* ν 9;

and the possible open chains are

5 λ 5, *5 6* μ 6 5, *5 6* 9 ν *9* 6 5,
10 ν *10*, 10 9 μ *9 10*, 10 9 *6* λ 6 *9 10*,
5 6 9 10,
5 λ *6* 9 10, 5 6 μ 9 10, 5 6 *9* ν 10,
5 λ 6 μ 9 10, *5* λ 6 *9* ν 10, *5 6* μ *9* ν 10,
5 λ *6* μ *9* ν 10,

and eight others which can be derived from the last eight of these by transposing Roman and italic numerals.

In all cases, the conditions imposed by the tetrads

(5, *5*, 9, *9*), (5, *5*, 10, *10*), (6, *6*, 10, *10*)

are automatically satisfied. Remembering that λ stands for **3** or **4**, μ for **7** or **8**, ν for **11** or **12**, we now have thirty-one reflexible icosahedra, whose faces are

0, 1, 2, 13, 3 4, 7 8, 11 12,
λ 6 *6* μ, λ 6 *6* 9 *9* ν, μ 9 *9* ν,
λ 5 *5*, 5 *5* 6 *6* μ, 5 *5* 6 *6* 9 *9* ν,
10 *10* ν, μ 9 *9* 10 *10*, λ 6 *6* 9 *9* 10 *10*,

and twenty-seven enantiomorphous pairs, whose faces (one from each pair) are

PLATE VI

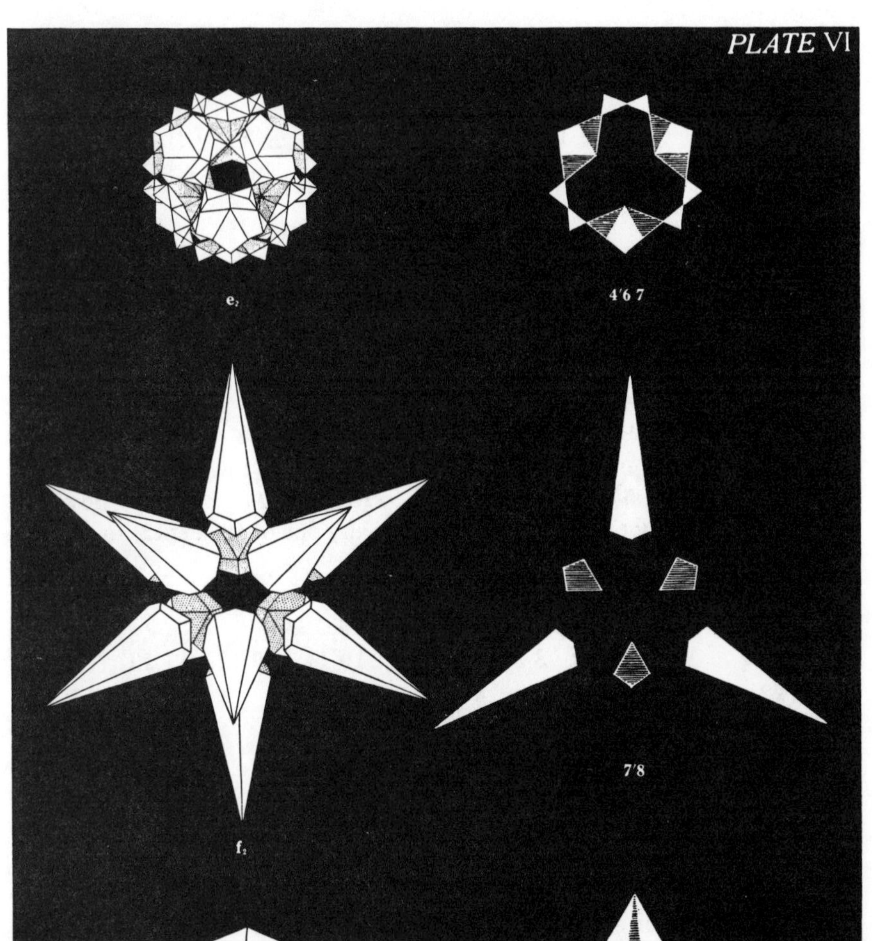

e_2 4′6 7

f_2 7′8

g_2 8′9′11

5 *6* 9 10,

λ 5 *6* 9 10, 5 6 μ 9 10, 5 6 *9* 10 ν ,

λ *5* 6 μ 9 10, λ *5* 6 *9* 10 ν , *5* *6* μ *9* 10 ν ,

λ 5 *6* μ *9* 10 ν .

The enantiomorphous pair *5* *6* 9 10, 5 6 *9* *10* is peculiar in having no common part. Accordingly, we add the combination

5 *5* 6 *6* 9 *9* 10 *10*

to the list of reflexibles. The enumeration is now complete.

It is convenient to simplify the symbols for the reflexible faces by writing **5** for 5 *5*, **6** for 6 *6*, **9** for 9 *9*, **10** for 10 *10*. We may now let λ stand for any one of **3**, **4**, **5**, and ν for any one of **10**, **11**, **12**, so that the list of (thirty-two) reflexibles becomes

0, **1**, **2**, **13**,

3 4, **3 5**, **4 5**, **7 8**, **10 11**, **10 12**, **11 12**,

λ **6** μ , μ **9** ν ,

λ **6 9** ν .

(For the chiral figures, we must let λ and ν retain their original meaning.)

One symmetry of the graph (Fig. 5) exhibits the duality between enantiomorphous pairs of solids. Another exhibits the possibility of subtracting every number from 15; but this duality has no obvious geometrical significance. (We may regard **2** as being paired with **13**, but **0** and **1** are exceptional in having no companions.)

So far, we have made no distinction between "positive" and "negative" regions of a face, *i.e.*, between regions where the inside ("substance") of the solid is *below* the plane of the face (regarded as lying above the centre) and regions where the inside is *above*. Hereafter, we shall distinguish a "negative" region by affixing a dash (′) to the number; *e.g.*, we shall write **3′ 5** instead of **3 5**. In the Plates, each solid is drawn on the left, and its face on the right (with "negative" regions shaded).

PLATE VII
e_2f_2
4'6 8
$e_2f_2g_2$
4'6 9'11
f_2g_2
7'9'11

3. AN ALTERNATIVE ENUMERATION, BY CONSIDERING SOLID CELLS

A somewhat different notation for the icosahedra, with perhaps a clearer idea of their character, can be obtained by considering, instead of the regions into which each plane is divided by the traces on it of the others, the analogous three-dimensional regions, or *cells*, into which space is divided by the whole set of twenty planes. We shall employ one clarendon symbol for a whole set of cells permuted into each other by the extended icosahedral group. Such a set may consist of 12, 20, 30, 60 or 120 cells, according to the symmetry of the individual cell; in the last case, the cell has no symmetry at all, and the set consists of two enantiomorphous sub-sets of 60 each, which will be indicated by corresponding Roman and italic symbols, as in the case of the plane regions. (This is found to occur in the case of one set of cells only, that forming the solid whose face-symbol is **5′ 6′ 9 10**; in all other cases the single region is reflexible.)

Any combination of these sets of cells forms a stellated icosahedron such as we seek, save that Miller's last two conditions (at the end of § 1) serve to rule out certain combinations; his other conditions are satisfied automatically. The fifth condition demands that the sets of cells in any admitted combination shall be *connected*, by having plane faces in common successively (*i.e.*, that the sets shall not fall into two groups, those of one group having no face in common with those of another). The fourth condition makes the same demand on the sets excluded from an admitted combination, counting amongst these the space external to the whole figure. This means that the face-symbol for the combination, derived from the face-symbols for the individual sets by cancelling out the dashed and undashed numerals representing faces in contact, must not include the whole face-symbol for any other solid, either in its proper form or with dashed or undashed numerals interchanged. (See the end of § 2.)

We classify the cells into a number of *layers*, each consisting of one or more sets, by the following principle. We define the *power* of any point, not lying in any of the twenty planes, as being the number of these planes that meet the rectilinear segment joining the given point to the centre, *i.e.*, the number of faces of the

PLATE VIII

De

4 5

Ef

7 9 10

Fg

8 9 12

icosahedron to which the point is external. Then the totality of points of given power forms a layer, bounded by portions of the planes, and thus consisting of some of the cells; consisting in fact of one or more complete sets of them, since it has the symmetry of the whole figure. These cells have edges but not faces in common, and form a spherical shell completely surrounding the centre.

The totality of points of power 0 we shall call **A**; it is the original icosahedron. The layers of power 1,2, . . . , 7, we shall call **b**,**c**, . . . , **h**, respectively, while the totality of points of power $\leqslant$ 1,2, . . . , 7, we shall call **B**,**C**, . . . , **H**, respectively. (There are also points of power 8,9,10; but their loci extend to infinity, and so are not layers of the kind we are seeking.)

The idea of the power of a point can be applied also to the plane face shown in Fig. 4; and it is clear that the solid consisting of all points of power $\leqslant n$ in space, is bounded by the plane figure consisting of all points of power n in Fig. 4. Thus we have, for the boundaries of the various solids of this type:

A	**B**	**C**	**D**	**E**	**F**	**G**	**H**
0	**1**	**2**	**3 4**	**5 6 7**	**8 9 10**	**11 12**	**13**

The various layers are given by subtracting **A** from **B**, **B** from **C**, and so on; accordingly, their boundaries are as follows:

b	**c**	**d**	**e**	**f**
0′ 1	**1′ 2**	**2′ 3 4**	**3′ 4′ 5 6 7**	**5′ 6′ 7′ 8 9 10**

g	**h**
8′ 9′ 10′ 11 12	**11′ 12′ 13**

Of these layers, it is obvious that all but three consist of a single set of cells only; these three however, namely **e**,**f**,**g**, each break up into two sets of cells:

$\mathbf{e_1}$	$\mathbf{e_2}$	$\mathbf{f_1}$	$\mathbf{f_2}$	$\mathbf{g_1}$	$\mathbf{g_2}$
3′ 5	**4′ 6 7**	**5′ 6′ 9 10**	**7′ 8**	**10′ 12**	**8′ 9′ 11**

Finally, $\mathbf{f_1}$ falls into two enantiomorphous subsets:

f_1	f_1
5′ 6′ 9 10	5′ 6′ *9 10*

(see Fig. 3).

De_1f_1

4 6′9 10

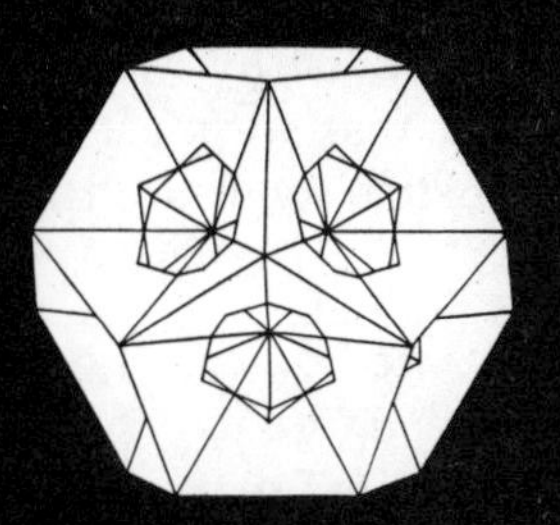

$De_1f_1g_1$

4 6′9 12

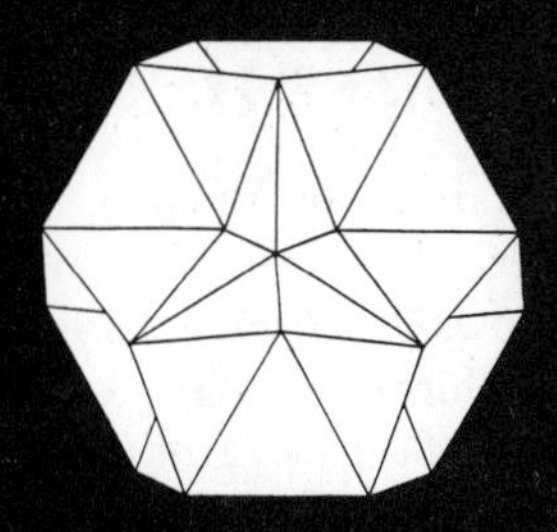

Ef_1g_1

7 9 12

The chiral solids are precisely those which contain one but not both of f_1, f_1.

It is clear that in combining these elements to form our solids, if the whole of one layer is present, every layer internal to it must also be present; and that if parts or the whole of more than one layer are present, these must be consecutive layers. Thus, as **e**,**f**,**g** are the only layers which break up into more than one set of cells, the reflexible solids (apart from the eight **A**,**B**, . . . , **H**) must have the forms

$$\begin{array}{cccccc} \mathbf{e}_i, & \mathbf{f}_i, & \mathbf{g}_i, & \mathbf{e}_i\mathbf{f}_j, & \mathbf{e}_i\mathbf{f}_j\mathbf{g}_k, & \mathbf{f}_j\mathbf{g}_k, \\ \mathbf{De}_i, & \mathbf{Ef}_i, & \mathbf{Fg}_i, & \mathbf{De}_i\mathbf{f}_j, & \mathbf{De}_i\mathbf{f}_j\mathbf{g}_k, & \mathbf{Ef}_j\mathbf{g}_k, \end{array}$$

where each of i,j,k is either 1 or 2, but not independently. In fact, since $\mathbf{f}_2$ has faces in common only with $\mathbf{e}_2$ and $\mathbf{g}_2$, the combinations $\mathbf{e}_1\mathbf{f}_2$, $\mathbf{e}_2\mathbf{f}_1$, $\mathbf{f}_1\mathbf{g}_2$, $\mathbf{f}_2\mathbf{g}_1$ are ruled out, the first and last by Miller's fifth condition, the other two by his fourth; so that we must have in every case $i = j = k$, and the possible reflexible solids are the eight **A**,**B**, . . ., **H**, and the following twenty-four:

$$\begin{array}{llllll} \mathbf{e}_1, & \mathbf{f}_1, & \mathbf{g}_1, & \mathbf{e}_1\mathbf{f}_1, & \mathbf{e}_1\mathbf{f}_1\mathbf{g}_1, & \mathbf{f}_1\mathbf{g}_1, \\ \mathbf{e}_2, & \mathbf{f}_2, & \mathbf{g}_2, & \mathbf{e}_2\mathbf{f}_2, & \mathbf{e}_2\mathbf{f}_2\mathbf{g}_2, & \mathbf{f}_2\mathbf{g}_2, \\ \mathbf{De}_1, & \mathbf{Ef}_1, & \mathbf{Fg}_1, & \mathbf{De}_1\mathbf{f}_1, & \mathbf{De}_1\mathbf{f}_1\mathbf{g}_1, & \mathbf{Ef}_1\mathbf{g}_1, \\ \mathbf{De}_2, & \mathbf{Ef}_2, & \mathbf{Fg}_2, & \mathbf{De}_2\mathbf{f}_2, & \mathbf{De}_2\mathbf{f}_2\mathbf{g}_2, & \mathbf{Ef}_2\mathbf{g}_2. \end{array}$$

The face-symbols for these twenty-four are respectively:

3′ 5,	**5′ 6′ 9 10,**	**10′ 12,**	**3′ 6′ 9 10,**	**3′ 6′ 9 12,**	**5′ 6′ 9 12,**
4′ 6 7,	**7′ 8,**	**8′ 9′ 11,**	**4′ 6 8,**	**4′ 6 9′ 11,**	**7′ 9′ 11,**
4 5,	**7 9 10,**	**8 9 12,**	**4 6′ 9 10,**	**4 6′ 9 12,**	**7 9 12,**
3 6 7,	**5 6 8,**	**10 11,**	**3 6 8,**	**3 6 9′ 11,**	**5 6 9′ 11.**

With regard to the chiral solids, our arrangement must be slightly different. We shall, for precision, consider that form of each which contains f_1 but not f_1; this we shall call the *dextro* form, and the other the *laevo*. The *dextro* form (of any of the chiral solids) has evidently the face-regions 5 or *5*′, according as the set of cells $\mathbf{e}_1$ is or is not part of it; 6 or *6*′, according as $\mathbf{e}_2$ is or is not part of it; *9*′ or 9, according as $\mathbf{g}_2$ is or is not part of it; and *10*′ or 10, according as $\mathbf{g}_1$ is or is not part of it. (The *laevo* form, of course, has either *5* or 5′, either *6* or 6′, *etc.*)

PLATE X
De_2
3 6 7
Ef_2
5 6 8
Fg_2
10 11

It is clear that f_1 (or f_1) can appear in combination with **E**, $\mathbf{e}_1$, $\mathbf{e}_2$, $\mathbf{De}_1$, $\mathbf{De}_2$, or with nothing interior to the layer **f**; while outside the layer **e**, it can be combined with nothing, $\mathfrak{g}_1$, $\mathfrak{g}_2$, $\mathbf{f}_2$, $\mathbf{f}_2\mathfrak{g}_1$ or $\mathbf{f}_2\mathfrak{g}_2$; but of the thirty-six combined possibilities which thus arise, nine are ruled out by the consideration that $\mathbf{f}_2$ must be present if both $\mathbf{e}_2$ and $\mathfrak{g}_2$ are, and cannot be present if neither of them is. The twenty-seven cases that remain are the following:

f_1,	$\mathbf{e}_1\ f_1$,	$\mathbf{D}\ \mathbf{e}_1\ f_1$,
$f_1\ \mathfrak{g}_1$,	$\mathbf{e}_1\ f_1\ \mathfrak{g}_1$,	$\mathbf{D}\ \mathbf{e}_1\ f_1\ \mathfrak{g}_1$,
$f_1\ \mathfrak{g}_2$,	$\mathbf{e}_1\ f_1\ \mathfrak{g}_2$,	$\mathbf{D}\ \mathbf{e}_1\ f_1\ \mathfrak{g}_2$,
$f_1\ \mathbf{f}_2\ \mathfrak{g}_2$,	$\mathbf{e}_1\ f_1\ \mathbf{f}_2\ \mathfrak{g}_2$,	$\mathbf{D}\ \mathbf{e}_1\ f_1\ \mathbf{f}_2\ \mathfrak{g}_2$,
$\mathbf{e}_2\ f_1$,	$\mathbf{D}\ \mathbf{e}_2\ f_1$,	$\mathbf{E}\ f_1$,
$\mathbf{e}_2\ f_1\ \mathfrak{g}_1$,	$\mathbf{D}\ \mathbf{e}_2\ f_1\ \mathfrak{g}_1$,	$\mathbf{E}\ f_1\ \mathfrak{g}_1$,
$\mathbf{e}_2\ f_1\ \mathbf{f}_2$,	$\mathbf{D}\ \mathbf{e}_2\ f_1\ \mathbf{f}_2$,	$\mathbf{E}\ f_1\ \mathbf{f}_2$,
$\mathbf{e}_2\ f_1\ \mathbf{f}_2\ \mathfrak{g}_1$,	$\mathbf{D}\ \mathbf{e}_2\ f_1\ \mathbf{f}_2\ \mathfrak{g}_1$,	$\mathbf{E}\ f_1\ \mathbf{f}_2\ \mathfrak{g}_1$,
$\mathbf{e}_2\ f_1\ \mathbf{f}_2\ \mathfrak{g}_2$,	$\mathbf{D}\ \mathbf{e}_2\ f_1\ \mathbf{f}_2\ \mathfrak{g}_2$,	$\mathbf{E}\ f_1\ \mathbf{f}_2\ \mathfrak{g}_2$.

The face-symbols for these are respectively:

$\mathit{5'}\ \mathit{6'}\ 9\ 10$,	$\mathbf{3'}\ 5\ \mathit{6'}\ 9\ 10$,	$\mathbf{4}\ 5\ \mathit{6'}\ 9\ 10$,
$\mathit{5'}\ \mathit{6'}\ 9\ \mathit{10'}\ \mathbf{12}$,	$\mathbf{3'}\ 5\ \mathit{6'}\ 9\ \mathit{10'}\ \mathbf{12}$,	$\mathbf{4}\ 5\ \mathit{6'}\ 9\ \mathit{10'}\ \mathbf{12}$,
$\mathit{5'}\ \mathit{6'}\ \mathbf{8'}\ \mathit{9'}\ 10\ \mathbf{11}$,	$\mathbf{3'}\ 5\ \mathit{6'}\ \mathbf{8'}\ \mathit{9'}\ 10\ \mathbf{11}$,	$\mathbf{4}\ 5\ \mathit{6'}\ \mathbf{8'}\ \mathit{9'}\ 10\ \mathbf{11}$,
$\mathit{5'}\ \mathit{6'}\ \mathbf{7'}\ \mathit{9'}\ 10\ \mathbf{11}$,	$\mathbf{3'}\ 5\ \mathit{6'}\ \mathbf{7'}\ \mathit{9'}\ 10\ \mathbf{11}$,	$\mathbf{4}\ 5\ \mathit{6'}\ \mathbf{7'}\ \mathit{9'}\ 10\ \mathbf{11}$,
$\mathbf{4'}\ \mathit{5'}\ 6\ \mathbf{7}\ 9\ 10$,	$\mathbf{3}\ \mathit{5'}\ 6\ \mathbf{7}\ 9\ 10$,	$5\ 6\ \mathbf{7}\ 9\ 10$,
$\mathbf{4'}\ \mathit{5'}\ 6\ \mathbf{7}\ 9\ \mathit{10'}\ \mathbf{12}$,	$\mathbf{3}\ \mathit{5'}\ 6\ \mathbf{7}\ 9\ \mathit{10'}\ \mathbf{12}$,	$5\ 6\ \mathbf{7}\ 9\ \mathit{10'}\ \mathbf{12}$,
$\mathbf{4'}\ \mathit{5'}\ 6\ \mathbf{8}\ 9\ 10$,	$\mathbf{3}\ \mathit{5'}\ 6\ \mathbf{8}\ 9\ 10$,	$5\ 6\ \mathbf{8}\ 9\ 10$,
$\mathbf{4'}\ \mathit{5'}\ 6\ \mathbf{8}\ 9\ \mathit{10'}\ \mathbf{12}$,	$\mathbf{3}\ \mathit{5'}\ 6\ \mathbf{8}\ 9\ \mathit{10'}\ \mathbf{12}$,	$5\ 6\ \mathbf{8}\ 9\ \mathit{10'}\ \mathbf{12}$,
$\mathbf{4'}\ \mathit{5'}\ 6\ \mathit{9'}\ 10\ \mathbf{11}$,	$\mathbf{3}\ \mathit{5'}\ 6\ \mathit{9'}\ 10\ \mathbf{11}$,	$5\ 6\ \mathit{9'}\ 10\ \mathbf{11}$.

We observe that fifteen of the twenty-seven chiral icosahedra are derivable from single reflexible icosahedra by adding f_1. In the remaining twelve cases, f_1 forms a "bridge" connecting two reflexible icosahedra (such as $\mathbf{Ef}_2$ and $\mathfrak{g}_1$) which would otherwise be disconnected (apart from contact along edges).

4. NOTES ON THE PLATES

PLATE I. The Icosahedron **A** needs no comment. **B** is obtained by erecting low triangular pyramids (**b**) on each face of

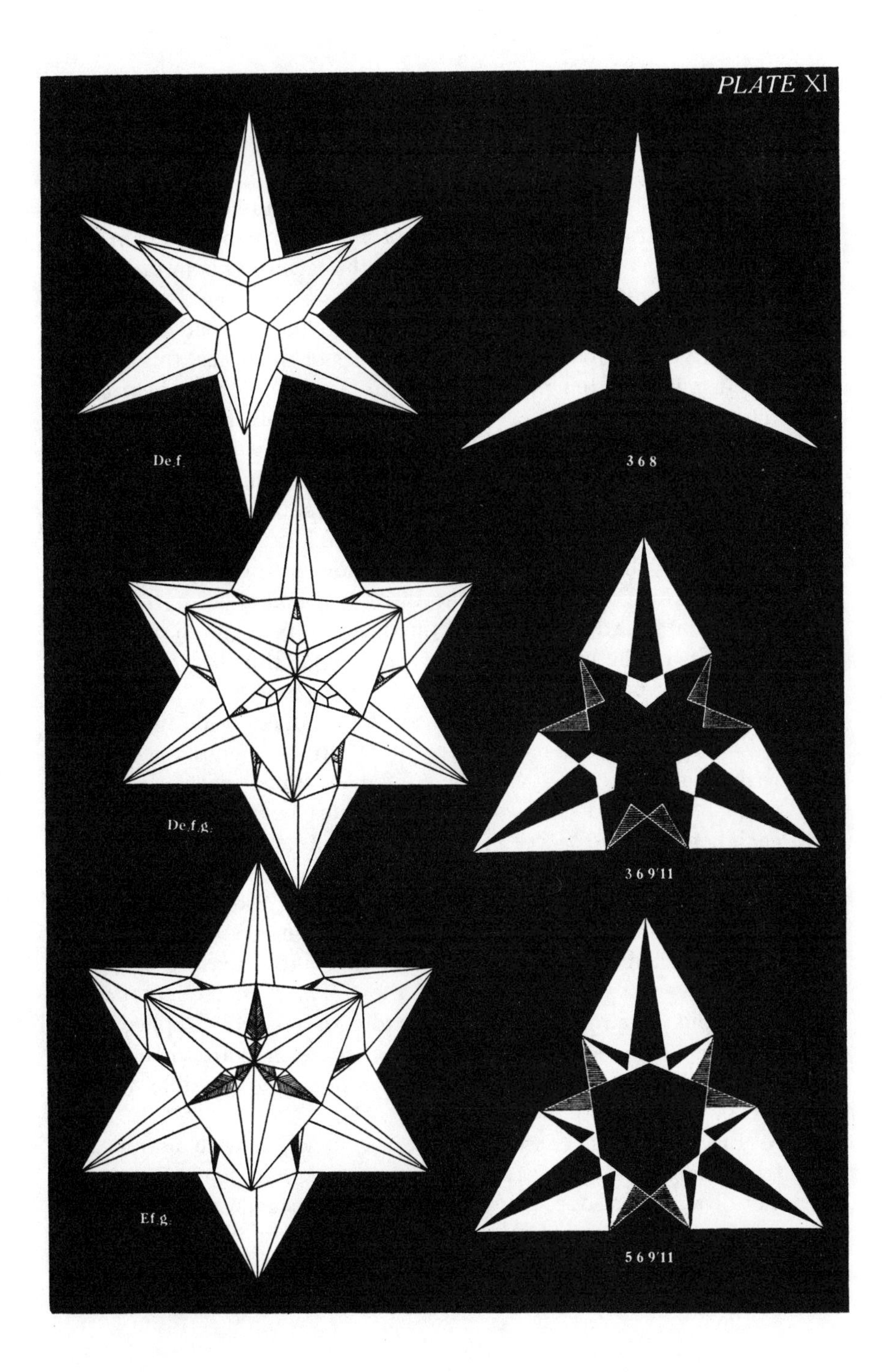
PLATE XI
De.f.
368
De.f.g.
369'11
Ef.g.
569'11

A, and is thus what we may call (by analogy with cristallographic terminology) a triakisicosahedron, save that the edges of **A** are concave, not convex, edges of **B**. **C** is the figure of five octahedra, dual to that of five cubes; each of the twenty planes contains a face each of two of these octahedra (the face of **C** is in fact clearly a pair of crossed triangles); and for each of the ten possible pairs out of the five octahedra a pair of opposite faces of one are coplanar respectively with a pair of opposite faces of the other, this accounting just for the ten pairs of opposite faces of the icosahedron. **D** has two sets of pits or depressions, 20 whose cross section (or vertex figure) is a triangle, and 12 of which it is a pentagram.

PLATE II. **E** is obtained by inserting into these trigonal depressions the bases of the ditrigonal spikes $\mathbf{e}_1$, and into the pentagrammatic depressions the inverted pyramidal sheaths $\mathbf{e}_2$; it thus has ditrigonal spikes and shallow pentagonal depressions. **F** is obtained by covering the ditrigonal spikes with the double pyramidal sheaths $\mathbf{f}_1$, and inserting into the pentagonal depressions the bases of the long pentagonal spikes $\mathbf{f}_2$; it thus has two kinds of spikes (long pentagonal and shorter double triangular), and rhombic depressions corresponding to those of $\mathbf{f}_1$. **G** is the familiar "great icosahedron", and is obtained by filling up the rhombic depressions with the wedges $\mathbf{g}_1$, and covering the spikes with the pentagrammatic sheaths $\mathbf{g}_2$; it has 60 isosceles triangular depressions, each between two pieces $\mathbf{g}_2$ and one piece $\mathbf{g}_1$.

PLATE III. **H** is obtained by inserting into each of these depressions a very long tapering spike, of isosceles triangular section; each spike has just one plane of symmetry. They fall into fairly marked clusters of five, and also into rings of six, their extreme vertices being very nearly those of a truncated icosahedron.

In the remaining plates some shading is generally applied to the negative or under sides of the planes, where they are visible in the remoter half of the figure; but it has not been possible to carry this out quite systematically.

PLATE IV. $\mathbf{e}_1$ consists of a set of 20 spikes with blunt bases, triangular below and fitting into the triangular depressions of **D**; and ditrigonal above, *i.e.*, the section is a hexagon whose sides are equal and whose angles have two values alternately, greater and less respectively than the angle of a regular hexagon. The set of

PLATE XII

f_1

5'6'9 10

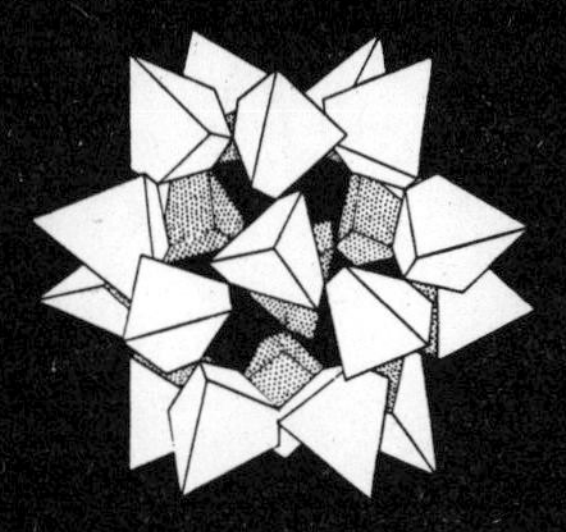

e_1f_1

3'5 6'9 10

De_1f_1

4 5 6'9 10

spikes is vertex-connected, *i.e.*, each has a vertex (but no edge) in common with each of three others. $\mathbf{f}_1$ consists of 120 scalene tetrahedra, and is the only one of the layers and partial layers whose individual pieces have no symmetry at all; 60 of them are of course the mirror images of the other 60. They fit together in cycles of six (three of each kind arranged alternately) to make 20 pyramidal sheaths, fitting over the spikes $\mathbf{e}_1$ and converting them into double pyramidal spikes, whose section consists of two crossed triangles, like the face of the five octahedra **C**. These sheaths meet by pairs in a couple of edges, so that the figure is edge-

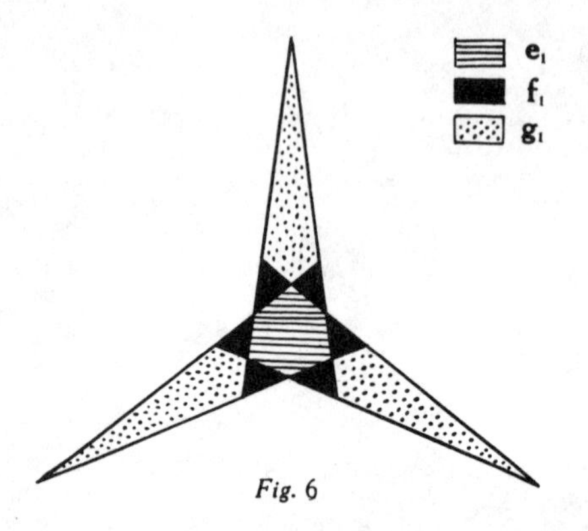

Fig. 6

connected throughout, forming a shell with twelve decagonal holes in it. Each double triangular spike has three wider and three narrower grooves down it, the former ending at the holes, and the latter meeting by pairs in 30 rhombic depressions. $\mathbf{g}_1$ consists of a set of narrow wedges which fit into these rhombic depressions, and whose acute edges form the edges of a large dodecahedron, with its vertices at the tips of the spikes $\mathbf{e}_1$. The accompanying sketch of the section of $\mathbf{e}_1$, $\mathbf{f}_1$, $\mathbf{g}_1$, by a plane perpendicular to a circumradius of this dodecahedron, and near its vertex, makes clear the relation between them. (In the plate, as in one or two others in which deep cavities are to be seen, the inner surfaces of the cavities of $\mathbf{f}_1$, into which the spikes $\mathbf{e}_1$ fit, are indicated by a somewhat heavier shading.)

PLATE V. The combinations of these three figures are easily grasped from the plate. $\mathbf{e}_1\mathbf{f}_1$ and $\mathbf{f}_1\mathbf{g}_1$ are both edge-connected, a

PLATE XIII

f_1g_1

5′6′9 10′12

$e_1f_1g_1$

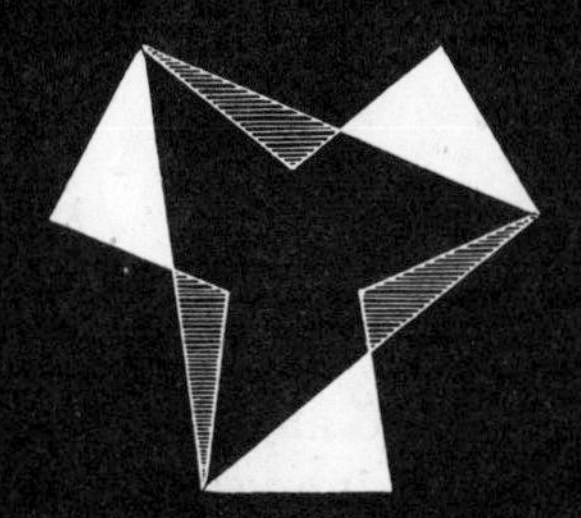

3′5 6′9 10′12

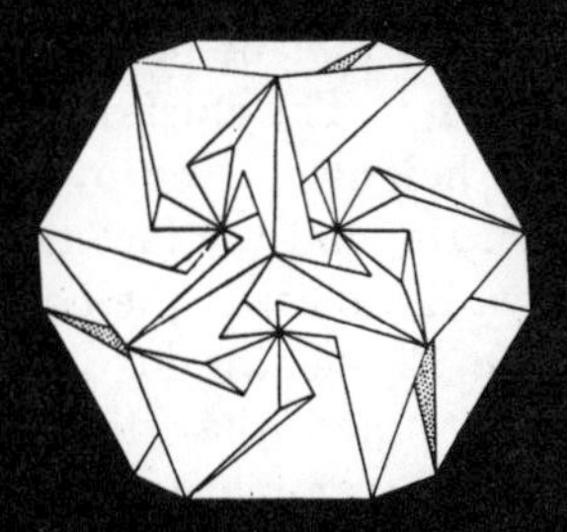

$De_1f_1g_1$

4 5 6′9 10′12

piece of the former consisting of a piece of $\mathbf{e}_1$ with six pieces of $\mathbf{f}_1$, a piece of the latter of a piece of $\mathbf{g}_1$ with four pieces of $\mathbf{f}_1$. $\mathbf{e}_1\mathbf{f}_1\mathbf{g}_1$ on the other hand is body connected throughout. All three form the same general type of shell with 12 decagonal holes.

PLATE VI. $\mathbf{f}_2$ is the only one of the figures whose pieces are not even vertex-connected. Each of the 12 is a long pentagonal spike with a comparatively blunt pentagonal base; the two pentagonal pyramids (acute and obtuse) by which it is bounded being oppositely placed, so that the edges of the one meet the faces of the other, and the lateral edges of each spike are a skew decagon, which just fits into one of the decagonal holes of $\mathbf{f}_1$, or of any of the figures in Plate V. The obtuse (inner) pyramid of each piece

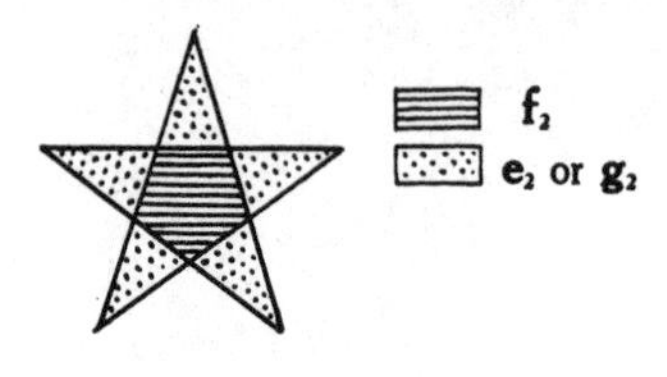

Fig. 7

is the solid angle vertically opposite to that of the original icosahedron **A** at a vertex. $\mathbf{e}_2$ and $\mathbf{g}_2$, though very dissimilar in appearance, are descriptively much alike; each is partly edge-connected and partly vertex-connected, the pieces fitting together by fives along edges to form 12 pyramidal sheaths, which fit onto one set of pyramids of $\mathbf{f}_2$, converting them from pentagonal to pentagrammatic pyramids. This is made clear by the accompanying sketch of a section of $\mathbf{f}_2$ and either $\mathbf{g}_2$ or $\mathbf{e}_2$ by a plane perpendicular to a pentagonal axis, and near a vertex of $\mathbf{f}_2$. The chief differences between $\mathbf{e}_2$ and $\mathbf{g}_2$ are the much greater height of each pyramidal sheath in the latter in proportion to its base, and the fact that the pyramids in the former are pointed inwards towards the centre of the figure, in the latter outwards. The 12 sheaths in each case are connected by vertices. It may be noted that the pentagrammatic under sides of $\mathbf{e}_2$ fit into the pentagrammatic depressions of **D**, converting them into pentagonal depressions. The five pieces of $\mathbf{e}_2$ at a pentagonal vertex occupy the solid angles vertically opposite to the five pieces of **B** at that vertex.

PLATE XIV

$f_1 g_2$

5′6′8′9′10 11

$e_1 f_1 g_2$

3′5 6′8′9′10 11

$De_1 f_1 g_2$

4 5 6′8′9′10 11

The interstices through which it is possible to see the inside of $\mathbf{g}_2$ are so narrow that it has been thought best to draw the figure in section, only (the farthest) 6 of the 12 sheaths appearing. The deep cavities which receive the spikes $\mathbf{f}_2$ are more heavily shaded, as in the case of $\mathbf{f}_1$.

PLATE VII. The combinations of these figures have the same kind of duality as $\mathbf{e}_2$ and $\mathbf{g}_2$ themselves. $\mathbf{e}_2\mathbf{f}_2$ and $\mathbf{f}_2\mathbf{g}_2$ each consist of 12 pieces, each bounded by a pentagonal and a pentagrammatic pyramid, with a skew decagon of lateral edges. The faces of the pentagonal are in each case indented in the base by the grooves of the pentagrammatic pyramid. The remaining solid has on each of its 12 pieces both the pentagrammatic pyramids and a sort of equatorial groove (decagonal, and following the lateral edges of $\mathbf{f}_2$) between the two sheaths. The 12 pieces of each solid are connected at vertices, in the case of $\mathbf{e}_2\mathbf{f}_2\mathbf{g}_2$ at vertices of two different kinds. (In the plate, $\mathbf{e}_2\mathbf{f}_2\mathbf{g}_2$ and $\mathbf{f}_2\mathbf{g}_2$ are shown in section, for the same reason as $\mathbf{g}_2$ in the previous plate.)

PLATE VIII. These three solids are very simple. $\mathbf{De}_1$ has the ditrigonal spikes of $\mathbf{e}_1$, their obtuser edges running down without interruption to the centres of the pentagrammatic depressions into which the pieces $\mathbf{e}_2$ fit; the figure can be regarded as consisting of 10 interpenetrating trigonal scalenohedra. $\mathbf{Ef}_1$ is the familiar figure of 10 interpenetrating tetrahedra, two inscribed to each of the five cubes, and two circumscribed to each of the five octahedra $\mathbf{C}$; in fact, the face is easily seen to consist, as far as its external outline is concerned, of two crossed triangles like the face of $\mathbf{C}$, but oppositely placed.

It may be noted that the indented faces of the pentagonal depressions exactly correspond to the inverted pyramids of $\mathbf{f}_2\mathbf{g}_2$ which fit into them; and that the rhombic depressions similarly correspond to the under sides of the narrow wedges $\mathbf{g}_1$.

$\mathbf{Fg}_1$ is most easily visualized as the result of fitting the pentagonal spikes $\mathbf{f}_2$ into the pentagonal depressions of $\mathbf{Ef}_1\mathbf{g}_1$ (Plate IX).

PLATE IX. Of these three figures $\mathbf{Ef}_1\mathbf{g}_1$ is the simplest; the rhombic depressions of $\mathbf{Ef}_1$ are just filled in level with the adjacent faces by the wedges $\mathbf{g}_1$, producing a figure which consists of a dodecahedron having a large pentagonal depression (with equilateral triangular sides) in each face. The other two, $\mathbf{De}_1\mathbf{f}_1$ and

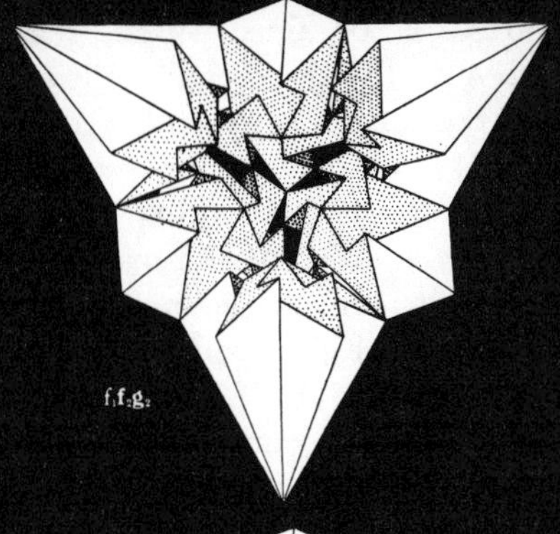

$f_1f_2g_2$

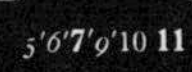

5′6′7′9′10 **11**

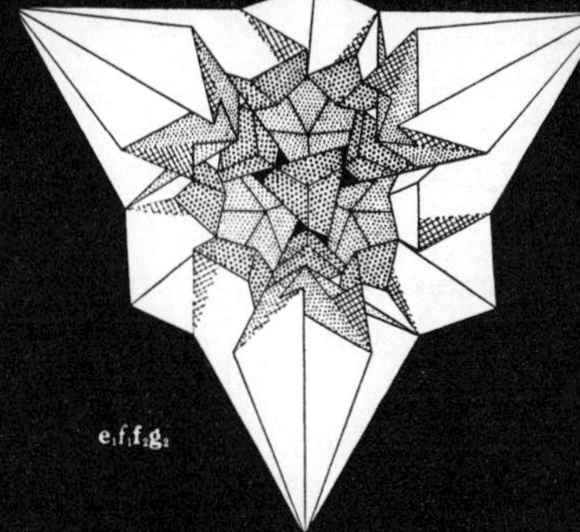

$e_1f_1f_2g_2$

3′5 6′7′9′10 **11**

$De_1f_1f_2g_2$

4 5 6′7′9′ 10 **11**

$\mathbf{De_1f_1g_1}$, are most simply thought of as obtained by the removal of $\mathbf{e_2}$ from $\mathbf{Ef_1}$ and $\mathbf{Ef_1g_1}$ respectively, the decagonal cavities and pentagrammatic depressions at the bottoms of them being a conspicuous feature of both.

PLATE X. The composition of $\mathbf{De_2}$ out of $\mathbf{D}$ and $\mathbf{e_2}$, as also its genesis by the removal of $\mathbf{e_1}$ from $\mathbf{E}$, are obvious on inspection of the figures. $\mathbf{Ef_2}$ is equally clear, the long pentagonal and short ditrigonal spikes belonging to $\mathbf{f_2}$ and $\mathbf{e_1}$ respectively. $\mathbf{Fg_2}$ can be thought of as obtained either by adding $\mathbf{f_2g_2}$ to $\mathbf{Ef_1}$, or by removing the wedges $\mathbf{g_1}$ from the great icosahedron $\mathbf{G}$; it may be regarded as consisting of six interpenetrating pentagrammatic scalenohedra, of which one is sketched below.

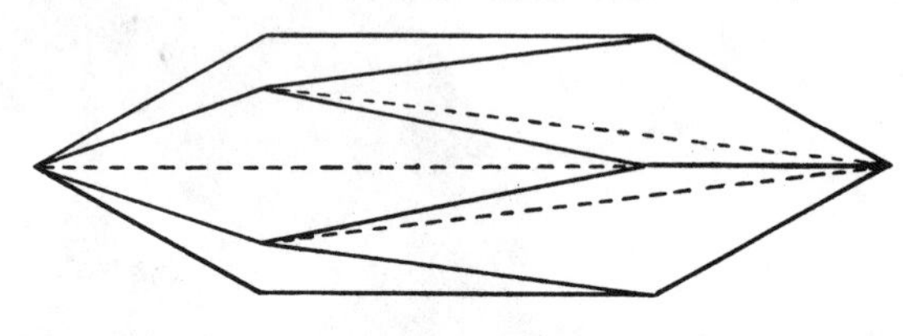

Fig. 8

PLATE XI. $\mathbf{De_2f_2}$ is a particularly simple figure, the spikes f_2 just fitting into the pentagonal depressions of $\mathbf{De_2}$ to form longer spikes which disappear into the centre of the solid; it can if we like be regarded as consisting of six interpenetrating figures, each of which consists of two oppositely placed pentagonal pyramids, and can be held to be the pentagonal analogue of the rhombohedron. $\mathbf{Ef_2g_2}$ is formed by fitting the inverted pentagonal pyramids $\mathbf{f_2g_2}$ into the pentagonal depressions of $\mathbf{E}$, which, however, receive only the portions next the vertices of these pyramids, their portions farther from the vertices being in contact only along edges with the more obtuse edges of the ditrigonal pyramids $\mathbf{e_1}$ (to which they are the vertically opposite dihedral angles). In the plate, the visible faces of $\mathbf{e_1}$ are shaded with rulings; it is seen that three pieces of $\mathbf{g_2}$ meet at a vertex over each of these pyramids, which are visible in the interstices between. $\mathbf{De_2f_2g_2}$ is obtained by omitting the spikes $\mathbf{e_1}$, leaving fairly large cavities under each of these meeting points of three pieces $\mathbf{g_2}$; the trigonal depressions at the bottoms of these cavities are more or less visible in the plate.

e_2f_1

$4'5'6\ 7\ 9\ 10$

De_2f_1

$3\ 5'6\ 7\ 9\ 10$

Ef_1

$5\ 6\ 7\ 9\ 10$

The remaining nine plates contain the chiral figures. Each of these exists in two enantiomorphous forms, one containing f_1 and the other f_1. In each case the figure containing f_1 (which we call *dextro*, the other being *laevo*) has been depicted.

PLATE XII. The most important figure on this plate is of course f_1 itself; it consists just of 60 out of the 120 pieces of $\mathbf{f}_1$, which are directly congruent to each other, and enantiomorphous to the rest; as each of the pieces which has an edge in common with any one is enantiomorphous to it, the set of 60 scalene tetrahedra is only vertex-connected, each piece having its most acute vertex in common with two others, and a different vertex in common with one other; it may be noted that at this latter vertex an edge of one is collinear with an edge of the other, the two faces at the one edge being respectively coplanar with those at the other.

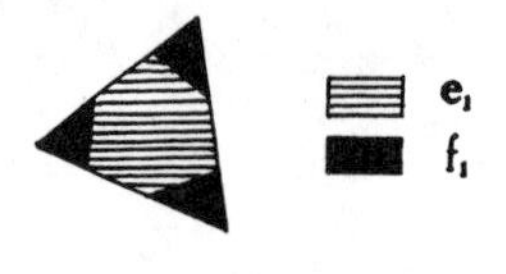

Fig. 9

These pieces combine with the ditrigonal spikes $\mathbf{e}_1$ to form triangular spikes, whose section near a vertex is shown in Fig. 9. (Cf. Fig. 6.) Of each of these spikes the outer and more acute triangular pyramid is set on obliquely to the inverted and more obtuse, forming something rather like a trigonal trapezohedron, save for the difference in acuteness.

Finally, the figure $\mathbf{De}_1\mathrm{f}_1$ is easily grasped, either by imagining the 60 pieces f_1 added to $\mathbf{De}_1$, or the 20 pieces $\mathbf{e}_1\mathrm{f}_1$ to $\mathbf{D}$, or (perhaps the most instructively) the pentagrammatic sheaths $\mathbf{e}_2$ removed from the simple and important figure of five tetrahedra, $\mathbf{E}\mathrm{f}_1$ (Plate XVI), leaving the characteristic pentagrammatic depressions.

PLATE XIII. The three solids $\mathrm{f}_1\mathbf{g}_1$, $\mathbf{e}_1\mathrm{f}_1\mathbf{g}_1$, $\mathbf{De}_1\mathrm{f}_1\mathbf{g}_1$, are formed in an obvious way by the addition of the wedges $\mathbf{g}_1$ to the last three. Each piece of $\mathbf{g}_1$ is in contact with two pieces f_1 and forms with them a sort of skew wedge, one of whose edges is formed of the collinear edges of the two pieces f_1. $\mathrm{f}_1\mathbf{g}_1$ consists just of these, connected only at their vertices; the addition of $\mathbf{e}_1$ to these makes

PLATE XVII

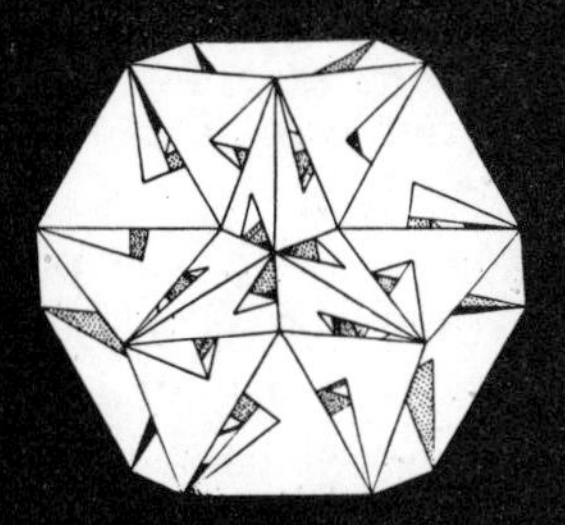

$\mathbf{e_2f_1g_1}$

4′*5′***6 7 9** *10′***12**

$\mathbf{De_2f_1g_1}$

3 *5′***6 7 9** *10′***12**

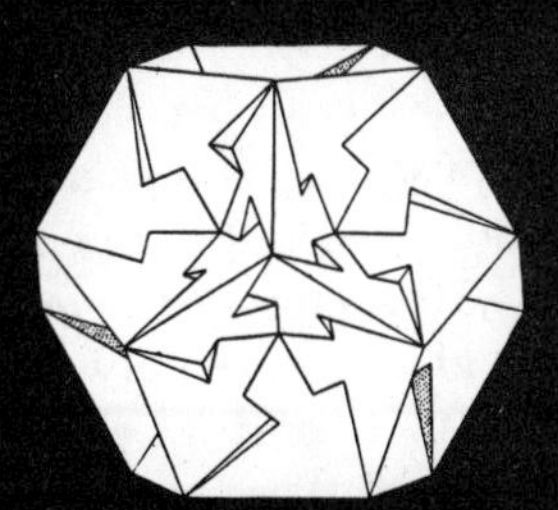

$\mathbf{Ef_1g_1}$

5 6 7 9 *10′***12**

the figure a body-connected shell with twelve holes; while of course $\mathbf{De}_1\mathrm{f}_1\mathbf{g}_1$ is a simply-connected solid. This last may perhaps most simply be thought of as arising by the removal of $\mathbf{e}_2$ from the comparatively simple figure $\mathbf{E}\mathrm{f}_1\mathbf{g}_1$.

PLATE XIV. Each piece $\mathbf{g}_2$ is in contact with one piece f_1, which fits onto its bevelled under side, and converts the latter into a single oblique plane. $\mathrm{f}_1\mathbf{g}_2$ accordingly consists of 12 pentagrammatic sheaths like $\mathbf{g}_2$ (of which as before only six are shown) except that each terminates towards the centre of the figure in a cyclic arrangement of five points, arranged rather as though to bore holes in wood. Of $\mathbf{e}_1\mathrm{f}_1\mathbf{g}_2$ likewise only half is shown, but as the addition of $\mathbf{e}_1$ makes the figure body-connected, it has been necessary to cut it in twelve places; these planes of section (which of course are not faces of the figure) are shown by rectangular cross rulings. $\mathbf{De}_1\mathrm{f}_1\mathbf{g}_2$ has large cavities inside (into which the spikes $\mathbf{e}_2\mathbf{f}_2$ would fit) communicating with outside space by chinks so narrow that it is quite impossible to see in; to make this clear, three pieces $\mathbf{g}_2$ and the pieces f_1 which are in contact with these are removed in the plate—if this were not done, the figure would be indistinguishable to the eye from $\mathbf{De}_1\mathrm{f}_1\mathbf{f}_2\mathbf{g}_2$.

PLATE XV. The relation of these three figures to the last three is obvious. As before, of the first two only half is shown, $\mathbf{e}_1\mathrm{f}_1\mathbf{f}_2\mathbf{g}_2$ having had to be cut in twelve places. In the case of $\mathbf{De}_1\mathrm{f}_1\mathbf{f}_2\mathbf{g}_2$ the external view is just what that for $\mathbf{De}_1\mathrm{f}_1\mathbf{g}_2$ would have been if part had not been cut away; it must be understood that the spikes are solid inside (owing to the presence of $\mathbf{f}_2$) but have deep chinks or crannies under their bases (owing to the absence of $\mathbf{e}_2$).

PLATE XVI. The combination $\mathbf{e}_2\mathbf{f}_1$ would of course be impossible, both halves of $\mathbf{f}_1$ being present, since it would form a complete shell; but with the omission of f_1 gaps appear which allow the inside of the solid $\mathbf{E}\mathrm{f}_1$ to be hollowed out. Each piece $\mathbf{e}_2$ is in contact by a face with one piece f_1, forming a scalene tetrahedron; and these are in contact by fives along edges only to form 12 (chiral) five-pointed stars; the 12 stars meet only at their vertices. In $\mathbf{De}_2\mathrm{f}_1$ the central pentagonal parts of these stars (namely $\mathbf{e}_2$) rest upon the solid core $\mathbf{D}$, but their rays meet by threes at vertices over cavities which would just hold the spokes $\mathbf{e}_1$. In $\mathbf{Ef}_1$

PLATE XVIII

these cavities are filled in, and we see that the trigonal vertices are precisely those of five interpenetrating tetrahedra; in fact, just as by adding the whole set $\mathbf{f}_1$ to $\mathbf{E}$ we obtain the whole figure of ten tetrahedra, by adding the half set f_1 we obtain five, one inscribed in each of the five cubes, and of course by adding the other half set f_1 we should obtain the other five. Either half set can in fact be defined as the total region interior to one set of five tetrahedra and exterior to the other, whereas $\mathbf{E}$ is the region interior to both. At the same time it is worth pointing out that the set of spikes $\mathbf{e}_1\mathrm{f}_1$ consists just of the regions interior to any one of this set of five tetrahedra and exterior to the other four, four pieces belonging to each tetrahedron; and of course $\mathbf{e}_1 f_1$ is related in the same way to the other set of five tetrahedra $\mathbf{E}f_1$.

PLATES XVII, XVIII, XIX. The addition of the wedges $\mathbf{g}_1$, of the spikes $\mathbf{f}_2$, and of both, to these three solids, is too clear from the plates to need further description.

PLATE XX. $\mathbf{e}_2\mathrm{f}_1\mathbf{f}_2\mathbf{g}_2$ consists of 12 body-connected pieces, connected with each other by vertices only, of which six are shown. The other two figures are quite clear in the plate.

PLATE XIX
$e_2 f_1 f_2 g_1$
4′5′6 8 9 10′12
$De_2 f_1 f_2 g_1$
3 5′6 8 9 10′12
$E f_1 f_2 g_1$
5 6 8 9 10′12

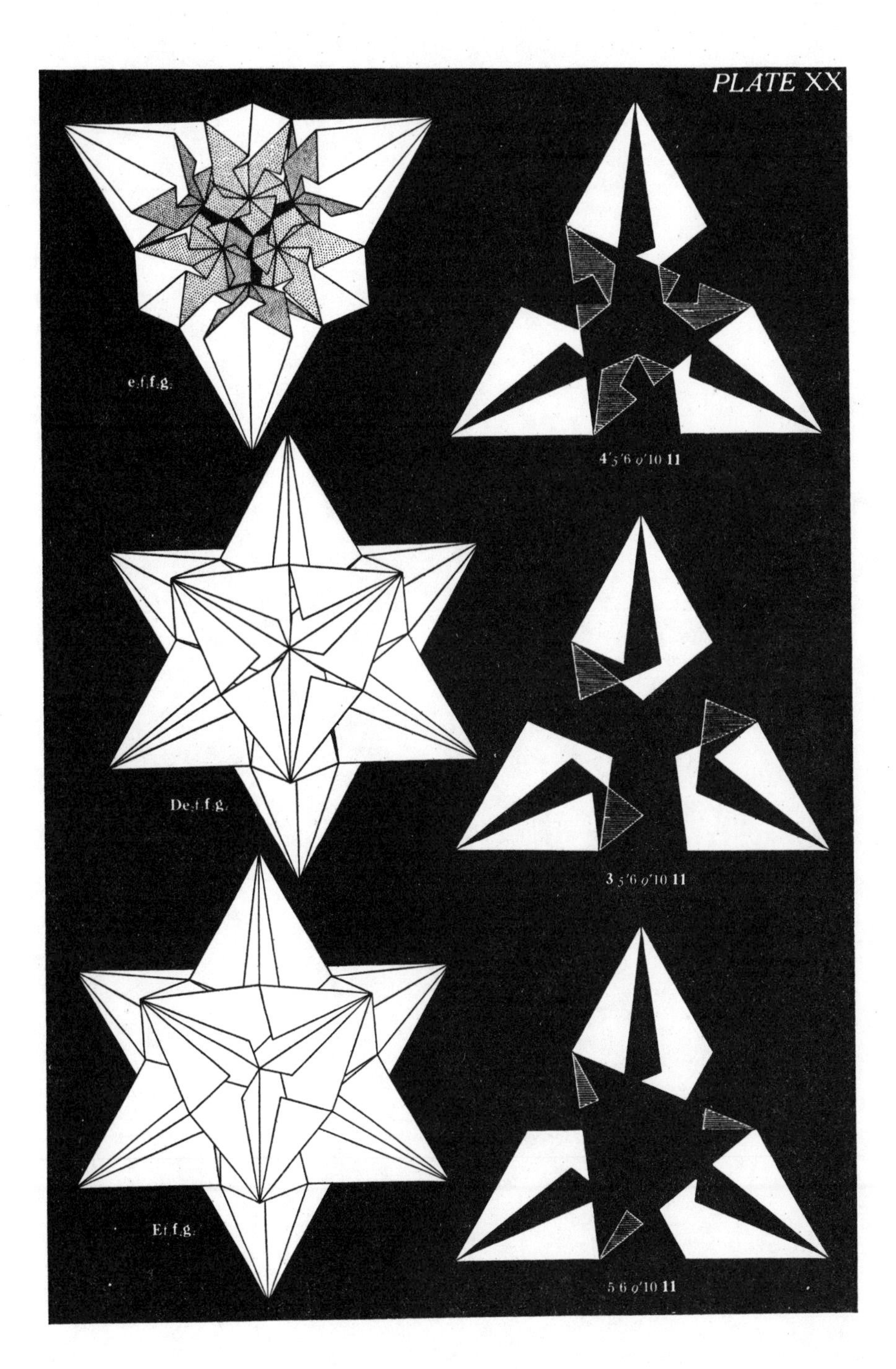
PLATE XX
e₁f₁f₂g₂
4′5′6 9′10 11
De₁f₁f₂g₂
3 5′6 9′10 11
E₁f₁g₁
5 6 9′10 11

The Geometric Vein

The Coxeter Festschrift

Edited by
Chandler Davis, University of Toronto
Branko Grünbaum, University of Washington
and **F.A. Sherk,** University of Toronto

Geometry's modern rebirth is explored through the underlying theme of H.S.M. Coxeter's work in this diverse collection of papers and survey articles. Though the scope is broad (encompassing group theory, combinatorics, and collateral fields), much of the material is presented in detail, and may be new even to specialists.

Of interest to research mathematicians as well as geometry teachers and advanced students, *The Geometric Vein* combines material of intuitive interest with a strong aesthetic appeal.

1981/viii, 598 pp./217 illus. including 5 color plates/Cloth/$48.00
ISBN 0-387-**90587**-1